PERPETUAL MOTION PHYSICS FOR NON-SKEPTICS

CONTENTS

Introductory Materials

- Devices that Don't Work —— p. 5
- Ignore the Sophisticates —— p. 13

Working Mechanics

- Is it Worth the Nobel? —— p. 17
- Partial Examples —— p. 21
- Nearly Complete Examples —— p. 29
- Experiments —— p. 39

V.E. PROJECT — p. 61

CALCULATIONS —— p. 63

BIO —————— p. 80

Nathan Coppedge

PERPETUAL MOTION PHYSICS

FOR NON-SKEPTICS

© 2015, '17, '18 Nathan Coppedge

Nathan Coppedge

INTRODUCTORY MATERIALS

- SOME DEVICES DON'T WORK-

You have probably heard before that there are perpetual motion machines that don't work. In spite of my optimism for some simple designs for perpetual motion, I admit that most of them do not work. But that does not mean that there are not workable principle in some part of these designs. Aspects like leverage, ramps, counterweights, and so on (even proportionality!) may be adopted in certain theoretically working designs, but they fail when applied to bad examples.

I will introduce working designs in later chapters. In this chapter, I am addressing designs that do not work. There is a lot to learn from them.

Nathan Coppedge

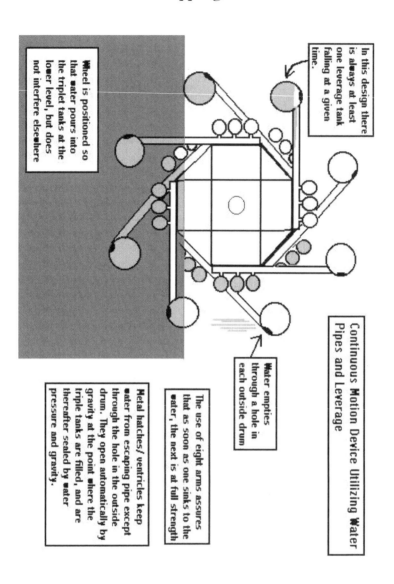

Continuous Motion Device Utilizing Water Pipes and Leverage

In this design there is always at least one leverage tank falling at a given time.

Wheel is positioned so that water pours into the triplet tanks at the lower level, but does not interfere elsewhere

Water empties through a hole in each outside drum

The use of eight arms assures that as soon as one sinks to the water, the next is at full strength

Metal hatches/ ventricles keep water from escaping pipe except through the hole in the outside drum. They open automatically by gravity at the point where the triplet tanks are filled, and are thereafter sealed by water pressure and gravity.

Depicted here is a Fluid Leverage type device. Like many examples from history, in particular, the Bhaskara Wheel designed around 1208 AD, this design runs into the difficulty of lifting half of its elements without a genuine change in mass distribution.

As engineers later learned, there was indeed no proportional advantage in these designs.

Later designs that I introduce have an advantage over this design through proportionality, solving the mass distribution problem, and through not needing to lift half of the elements, and sometimes, by introducing methods of cheating----not batteries, but some passive physical elements, or greater horizontal directionality.

Now I will show another device that does not work.

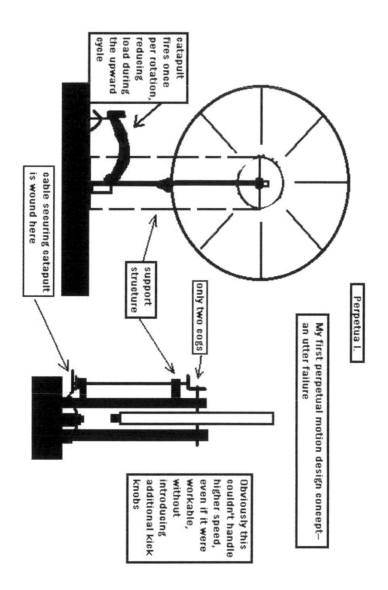

Devices like this make use of springs, and wear down quickly. Thus, they are not an option. In fact, upon circumspection, they seem impractical, because of the effects of friction. By the time the wheel spins 360 degrees to be hit again by the catapult, it has already slowed down, and does not have the force to wind up the catapult again to create the same effect. This is understandable, and in other designs we can be cautious to look out for this effect.

I argue, there may be other designs that are not so easily affected by friction, because they actually have an unbalanced force, e.g. through cheating without batteries.

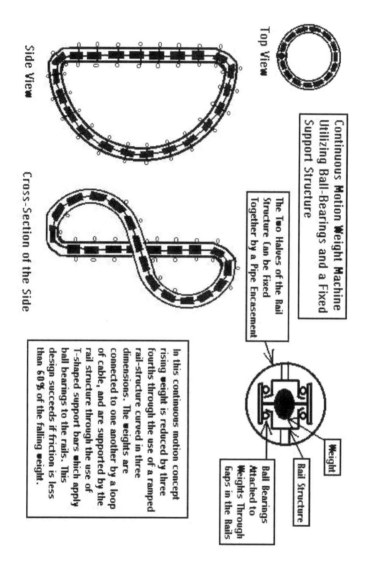

Another example appears to have an advantage, but appears to fail through proportionality. Perhaps if physics were very nice, with superfluid or the like it would be operable.

But, it must be assumed that because of the equal vertical distance between the top and bottom of the free-falling weights, and the top and bottom of the partly horizontal weights, that because the horizontal weights must be lifted, that it doesn't work. But, who knows, someone may try this and find out that it has an advantage after all.

At this point, we begin to find what may be called ambiguous cases. Or, at least, cases that might work for the creative imagination.

Nathan Coppedge

IGNORE THE SOPHISTICATES

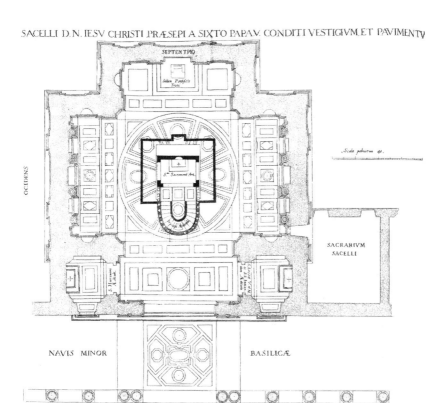

[credit: public domain]

Nathan Coppedge

You may think this is a design for a complex motor, running on magnets perhaps. But, in fact, it's not. It's a floor plan for the Sistene Chapel, a building from the Renaissance!

So, the lesson is, avoid unnecessary complexity! If you can't distinguish it from a diagram for the Sistene Chapel, then there's something very wrong! Probably, batteries are involved, or a hidden wire directing current!

Beware of the 'ersatz' 'complex' examples. Many of these designs using magnets make use of electricity, and therefore aren't real perpetual motion machines! And very often, their inventors won't even admit it! Probably, they're frauds! Certainly they are while they take your money.

THE MESSAGE: So, that's the message: avoid designs that have been proven not to work, and avoid designs that are over-complicated. Both of those cases are follies.

WORKING MECHANICS

Nathan Coppedge

PERPETUAL MOTION PHYSICS FOR NON-SKEPTICS

IS IT WORTH THE NOBEL PRIZE?

VIDEOS:
VERTICAL LEVER 1A1
VERTICAL LEVER 2B

Around July 2018, I published videos for the Vertical Lever design, whose ratios seemed extremely workable. However, it was somewhat difficult to work out the equations at first. There was so much skepticism circulating that I had trouble measuring properly.

Anyway, using a 3/4 mostly vertical lever, supported in one slot by a <22.5 degree track, and falling somewhat horizontally in the larger slot, the falling ratio is 3 versus 1.25, while the rising ratio is the same 1.25 versus 0.5+, comparing effective leverage on both ends, as the marble when supported has it's mass moving horizontally not vertically.

The counterweight does have a lot of mass, but it has less effective leverage, which my math teacher taught me still makes a difference. I asked him to tell me everything he knew about leverage , and it took him three weeks as I recall. I lapped it up.

If there is friction in the 3/4 Vertical Lever, the counterweight will be maximized around 7X to < 12 X, as 12X is where 12 X 0.25 = 3 resistance is met for the 1X mass and 3X leverage on the long end.

A complete description of the Vertical Lever:

1. A square or rectangular track structure set at a significantly less than 22.5 degrees with two longish rectangular slots, each about 4—6 inches long. The larger one should be at least twice as wide to permit the basket or fork to free-fall, attached to the lever from below the slots.
2. Short straight barriers or walls set around both tracks at least up to 3/4 of the marble at the relevant point, but with some spacing on the smaller track to allow the relevant size marble to move freely. There may be some slightly flexibility especially with larger balls and marbles. The two sots should be very narrowly separated with only one wall between.
3. A deflecting board should be used, slightly longer than the width of the walls of the smaller slot, and directed into the top of the larger slot.
4. The change in height of the entire length of the slots should be slightly more extreme than the variation in the arc of the end of the lever where the basket is located. This permits deflection into the basket from a slightly higher altitude at the top, and return into the narrower slot at the base. The fourth feature may place restrictions on the overall length of the slots relative to the length of the lever.
5. The ratio of the lever shall be 3/4 or 3:1, the long end being the end operated by the marble.

PERPETUAL MOTION PHYSICS FOR NON-SKEPTICS

6. The weight ratio shall always use whatever marble is used as the standard of measurement. I have found ideal values tend to place the counterweight at barely >7X to definitely less than 12X marble mass. Earlier calculations assumed a median rather than an average was desirable, and so the ideal number is located higher than before.
7. The pivot of the lever is to be located beneath and not very far from the top end of the track. 20 inches length seems to be the most workable number for serious experiments. The lever should be nearly vertically disposed, meaning the pivot point is deep under the track, but somewhat outside it to create the angle. The angle should not be perfectly vertical, but rather still leans somewhat to the side at it's highest motion, perhaps at least 6 degrees. FINALLY, if the sideways transition between tracks is fairly narrow compared to the overall scale and size of the marble, transitions should be fairly easy to facilitate.

One way to construct it is to use a largish ball and have the larger slot the same size as the walls of the smaller slot, but with the slot being as wide as the walls of the smaller slot. This allows some exciting standardization.

Perhaps this design is one of the easier perpetual motion machines to build, and I hope in the months following July 2018 I will get some 3-d printers including Jer Ram who I am working with who are willing to model this design.

Nathan Coppedge

PERPETUAL MOTION PHYSICS FOR NON-SKEPTICS

-PARTIAL EXAMPLES-

Here are some partial examples that may prove the principle of perpetual motion.

They are cases ostensibly of 'over-unity without the batteries'.

1. Here is an example that doesn't really work, but it illustrates the point.

Dominoes can *accelerate*.

[Credit: Nathan Coppedge]

Therefore, if dominoes don't need to be reset, then the result might be a perpetual motion machine.

2. Lifting a weight with an equal weight.

This is possible when the weight being pulled is supported by a slight upward incline. The pulling weight is unsupported, applying its weight to pull the other weight. Although the free-falling weight moves more vertical distance, the two weights move the same absolute distance. And, the free-falling weight can be located at a higher altitude, meaning that it did not move to a lower absolute altitude. Therefore, it can still interact with the weight that was moved, or the effects of the weight that was moved.

And, it should be emphasized: *the weight moved*! That's the bottom line in perpetual motion.

Here's an example that I believe uses this effect (the downwards cord at A is attached to the first see-saw). Ball weight A. moves the same distance as ball-weight D. The upper ends of see-saws B. and E. are downwards-sloped slightly.

PERPETUAL MOTION PHYSICS FOR NON-SKEPTICS

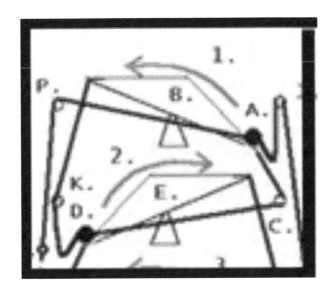

3. Moving Horizontally

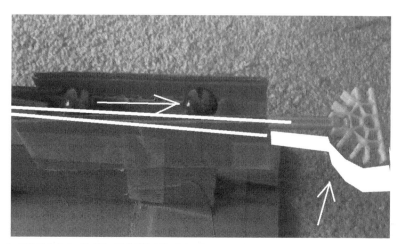

A SMALL DIP IN THE BEGINNING OF THE LEVER MAY BE ADEQUATE TO CONTINUE THE MOTION IN A LOOP. LET ME DEMONSTRATE

In the above diagram we see that the counterwight (not pictured on the left) has the force to *move the marble mostly horizontally, even when the two masses are equal, except accounting for the mass of the lightweight lever.*

This is even proven in experiment.

So, we get upwards and downwards motion at no inherent cost of energy. Perhaps altitude is not even lost.

4. Principled Asymmetry.

I found that certain arrangements of construction toys were more prone to move in one directly than another, without a stop. Thus, possibly over-unity.

Principled Asymmetry LOOKS unbalanced.

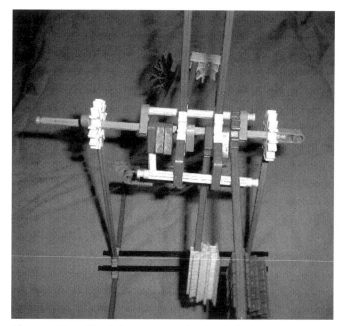

Above: Detail of the gears of principled asymmetry.

PERPETUAL MOTION PHYSICS FOR NON-SKEPTICS

5. In the case of an offset lever pendulum, forcefully turning the central point 180 degrees creates a 360 degree turn afterwards.

This, I believe, is against conventional physics.

If you study physics, you know this offset lever has relatively unique properties for a pendulum, but only when the string on which the weight dangles is at a certain length!

Nathan Coppedge

-NEARLY COM-
PLETE EXAMPLES-

I have made a list of examples which are nearly proven...

[Surprisingly I'm NOT talking about Reidar Finsrud's 'PM 1'.

Instead, I am talking about designs using simple mechanics that just might break the laws of thermodynamics, if not Newtonian Mechanics.]

HERE IS A SHORTLIST OF PERPETUAL MOTION CONCEPTS WHICH HAVE ONE REMAINING DESIGN PROBLEM, HIGHLY -SPECIFIC PROBLEMS I BELIEVE TO BE SOLVABLE:

The plain truth is that there are not many designs I have found that have only one thing remaining. I will try to make a list here with a list of the remaining problems, so as to give you [in this case, anyone] options to select from:

Single-Module First Fully Provable: in the original configuration (for example, similar to the recent Oddity video), which perhaps merely requires a tall weight because in the correct build it has a proven height advantage with room to move into.

PERPETUAL MOTION PHYSICS FOR NON-SKEPTICS

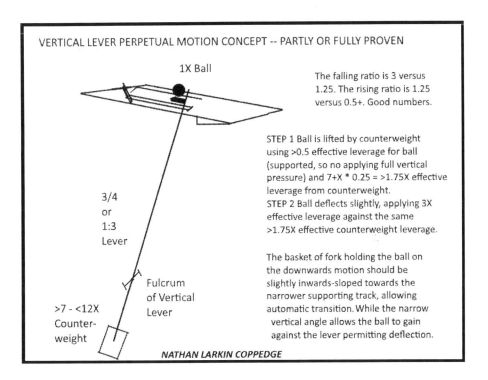

NATHAN LARKIN COPPEDGE

The Vertical Lever: has proven back-and-forth motion like the Not-If-But-When 4, (but with a different design) but merely needs to have enough angularity in how the ball is lifted so that the ball can be deflected onto a basket on the lever which can hold the ball temporarily and then roll off to be lifted again when not supported by the basket.

SUCCESSFUL PERPETUAL MOTION EXPERIMENT 4

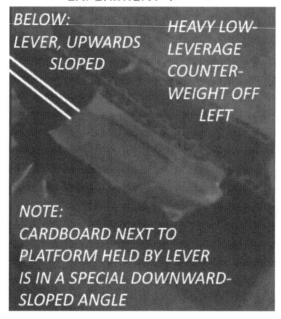

RIGHTWARD MOTION AT NO COST...

The Successful Perpetual Motion Experiment 4. There are strong indicators of what must be upward motion simply using downward angularity in a fixed side element. The ratios are tricky (the slightly vertical direction of the lever is very counter-intuitive), but this seems to be a proven principle for creating momentum at no cost that could work in many configurations, for example, working with short-distance momentum around a curve and an identical unit on the return. The challenge here is how to design the downward angularity of the fixed side element, but there is a video that can be analyzed to find out.

The Real Perpetual Motion Experiment 2: has some nifty properties. It is a bit involved, but I believe I discovered a principle that might be adapted to something similar to the Spiral Cone device creating more efficiency, so that the horizontal inward spiral assists the motion of the ball, but a significant part of weight is allowed to be applied on the downward, outward-directed return angle. You see, the inward spiral might assist even when the slope remains sharp. This is especially true with long upward angles that allow a gentle upward slope in the track but achieve high altitude. When altitude gain is maximized relative to the slope, the inward spiral can be extreme and still maintain a steep downwards drop on return. Adding to this, the downwards slope can be directed towards the lower beginning point of the slope (even using a projecting ramp!), so that the downwards slope creates an automatic return to the beginning position. The nifty nifty thing is that a counterweight is not illegal with this as long as its effective leverage is less than the effective leverage of the ball or wheel when the ball or wheel is unsupported. It seems to me the inward spiral can be made rather sharp in certain builds while maintaining a sharp downward angle on return, so it seems to me it would have natural momentum. In this case the materials might be rather flexible but the string for the pendulum would have to pivot easily in 360 degrees, which can impact design.

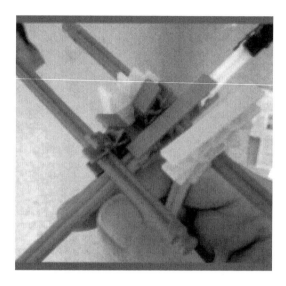

The Natural Torque Device: has proven torque and apparently the ability to gain net altitude in very precise configurations. If the exact 3-d angle of the Natural Torque Device that permits net upward motion could be found, it could then be joined to an upward spiral and many other applications. The easiest of these would hopefully be some type of lightweight turntable attachment. Given that it has proven torque and there is evidence of a very narrow window with net upward motion, making a fully-functional Natural Torque Device should not be impossible, but it fits under problems of degree.

THE ESCHER MACHINE

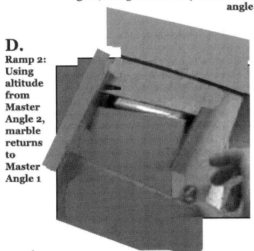

C. Master Angle 2: marble rolls upwards again, using a differently-directed master angle

D. Ramp 2: Using altitude from Master Angle 2, marble returns to Master Angle 1

B. Ramp 1: A downwards motion is possible due to the gain in height

A. Master Angle 1: marble rolls upwards using a horizontal slope

NATHAN COPPEDGE

The Escher Machine: has a ridiculously precise ratio, but there is one video that shows it working. The advantage here is that with a ridiculously precise ratio it could be built as one solid piece that would take many sizes of ball (usually with their midpoint about 33% - 50% above the lip of the track I have found).

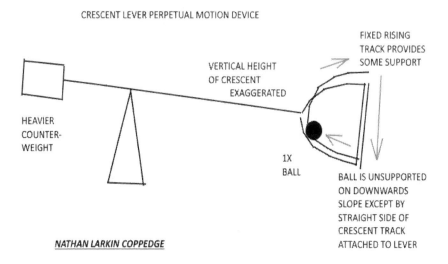

The Crescent Lever: is a rather efficient version of the Repeat Lever 2, the problem being that it needs to gain more altitude than the variation in the downwards-sloped drop point. I was astonished at this design, because its advantages seem over-the-top. It is sort of too-good-to-be-true that it is forced to apply leverage when it already has a downwards slope. This design could be very promising, and also easy to build.

PERPETUAL MOTION PHYSICS FOR NON-SKEPTICS

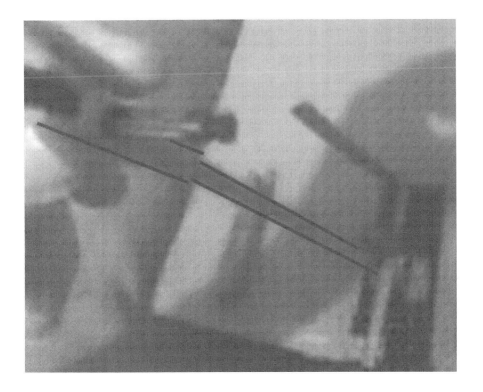

The Not-If-But-When 4: This has similar functionality to the Vertical Lever. Once gain, a somewhat upwards-directed angle would have to be produced in the track in the direction of the initial motion in order to permit gain against the initial altitude of the lever. The specific means of transition has to be worked out, but it appears to work in some configurations given that it was able to deflect off a backboard and return at a similar altitude including gains and loss.

A variation of the Coquette with supporting side track: This design is almost indistinguishable from some of the other designs, but it is a bit intriguing. Where the Not-If-But-When involving the circular wire only loses altitude at one point, the Coquette with support could have support for half of the circular motion and then use the downward drop to change the angle of the track from upwards to slightly downwards. This would allow the ball to press down on a slightly-sloped wire attached to the pivoting circle, and also later reach a lower position where the wire could be used to lift the ball slightly along a fixed track protruding from below into the circle until the ball is no longer supported. An advantage of this design which may be unnecessary is that the length of the wire versus the length of the downwards slope is completely flexible, they simply have to add up to the same change in altitude. So, this device is very similar to one of the Not-If-But-Whens and a number of the Repeat Lever designs.

EXPERIMENTS

Nathan Coppedge

PERPETUAL MOTION PHYSICS FOR NON-SKEPTICS

1. Tilt Motor Experiment

ABOVE: Wheel in initial position before rolling.

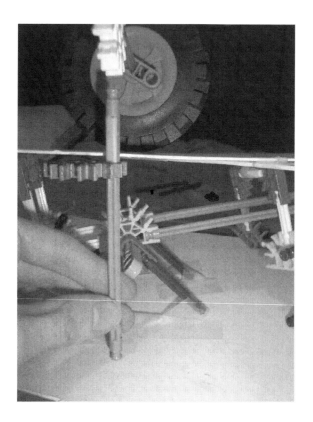

[Measuring before and afterwards seemed to show that the altitude had not changed, in spite of momentum].

[Top view of the experiment].

However, later experiments seemed disconfirmative of this design, or at least of the experimental method.

2. Motive Mass Experiments

I proved that a free-falling weight could lift a supported weight around 2007.

Here are the photographs.

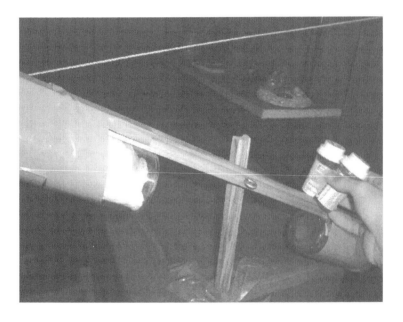

4 oz. applied at about 4 - 5 inches.

PERPETUAL MOTION PHYSICS FOR NON-SKEPTICS

The 4 oz. cart was lifted some vertical distance when pulled horizontally.

3. Principled Asymmetry.

Already mentioned, the plastic contruction appeared to move more easily in one direction than another by sheer weight imbalance, and appeared to sometimes stop in unbalanced positions.

These experiments date from 2007 - 2009.

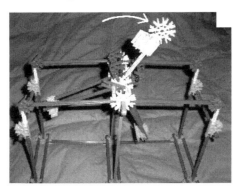

4. First Major Cheating Principle

Lifting a weight that is supported produced a clear advantage.

Believe it or not, the experiment is from as late as 2009 - 2014.

ABOVE: A surrounded connecting board is able to move a marble up a beveled slot.

5. Modular Trough Lever Experiment
['Successful Over-Unity Experiment 1']
From Nov. 9th—10th 2013

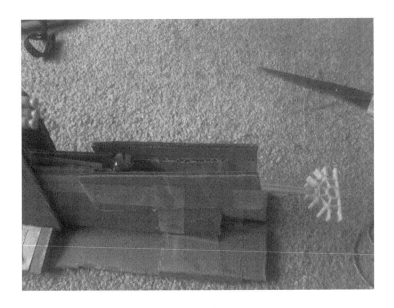

[Image 1: Marble rolls upwards slightly by force of the counterweight].

PERPETUAL MOTION PHYSICS FOR NON-SKEPTICS

[Image 2: Marble is able to fully lift counter-weight after movement].

SUCCESSFUL EXPERIMENT!

6. Escher Machine Experiments

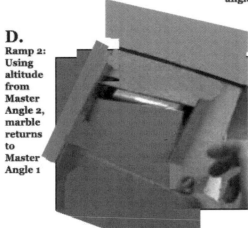

THE ESCHER MACHINE

C. Master Angle 2: marble rolls upwards again, using a differently-directed master angle

D. Ramp 2: Using altitude from Master Angle 2, marble returns to Master Angle 1

B. Ramp 1: A downwards motion is possible due to the gain in height

A. Master Angle 1: marble rolls upwards using a horizontal slope

NATHAN COPPEDGE

A certain arrangement seemed to show proneness to motion in four directions, all connected, appearing to show at least a subtle indication towards a 'Escher-Machine' type principle. Other types of experiments dealing with only one track were more confirmative [See videos by searching for 'master angle experiment' or 'escher perpetual motion machine' on Youtube or Google]

PERPETUAL MOTION PHYSICS FOR NON-SKEPTICS

7. Spiral Pendulum Experiment

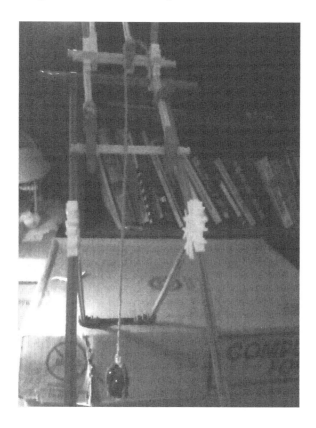

[Image 1: Spiral Pendulum from the side].

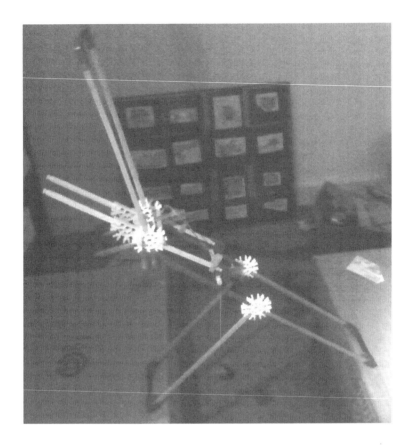

[Image 2: Spiral pendulum (Front View)].

The pendulum was able to swing 420 times, instead of the predicted 165 times.

PERPETUAL MOTION PHYSICS FOR NON-SKEPTICS

8. Over-Unity Lever Experiment
First Photo: the upwards bars are connected by a spacer and lightly weighted at the end (not pictured).

In this photo the bars have been pulled slightly downwards along an upward-inclined track (bottom), to which the lever (weighted bars) is connected through looped connectors.

When I let go, however, the lever moves UP the fulcrum due to an angled bar (shown in the photo).

You can see that the level shows an upward movement along the lower fulcrum.

53

9. Successful Partial Perpetual Motion Experiment 2

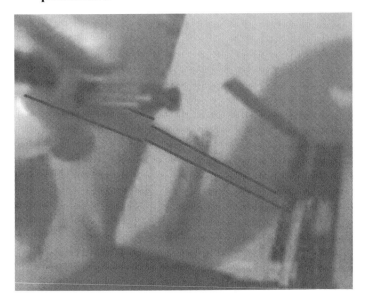

The counter-weight (left, not pictured) in this steep lever arrangement causes marble (barely visible, towards tip of lever) to move outwards when supported as shown, and the marble has enough weight to return the lever at the same altitude in the larger slot (right), in both cases with momentum.

Although the experiment was not initially able to create solid transitions, this may be seen as the first solid proof of the possibility of perpetual motion.

PERPETUAL MOTION PHYSICS FOR NON-SKEPTICS

10. Natural Torque Device

ABOVE: Natural Torque Device has proven torque

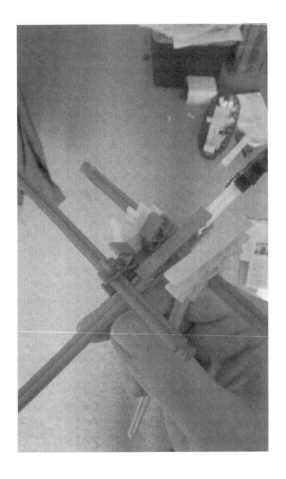

ABOVE: View of Natural Torque mechanism.

PERPETUAL MOTION PHYSICS FOR NON-SKEPTICS

ABOVE: View of the 'white wheel' of the Natural Torque Device, which has a proneness to move up a slope when the holding stick is bent towards the wheel.

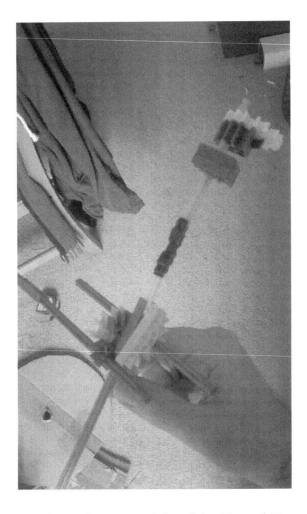

ABOVE: Counterweight of the Natural Torque Device.

11. SINGLE-MODULE FIRST FULLY PROVABLE EXPERIMENTS

ABOVE: The ball is able to be propelled left and right automatically once the lever is depressed.

ABOVE: The ball seems s to have a strong advantage at the beginning transition due to the height of the ball. However, horizontal motion may be limited to a few inches, including upwards and downwards motion. Weight ratios must be adjusted to accommodate the larger ball, necessitating a sturdier track.

INFLUENCE ON THE VISUAL EDUCATION PROJECT

At least two of my designs made it into the demonstrations by the famous Visual Education (VE) Project on YouTube:

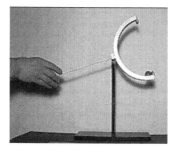

The Ball-and-Pendulum Design

The So-Called "Carousel Machine"

At the point that I suggested these two dysfunctional designs, Visual Education Project was already famous.

Nathan Coppedge

PERPETUAL MOTION PHYSICS FOR NON-SKEPTICS

EQUATIONS

A LIST OF PERPETUAL MOTION EQUATIONS

GENERAL PERPETUAL MOTION

Volitional Energy = Mobile U / Necessary Dual-Axial U

Volitional Equilibrium = Modular U / (Stems per cycle / subcycles per cycle)

Volitional Efficiency = Volitional Energy / Volitional Equilibrium

The vertical begins at 45 degrees.

Official General Theory: I estimate the barriers are 1, 1.5, 2, >2. At unity, energy can continue with zero resistance. At 1.5 advantages may be had. For example, 1.5X leverage, 1.5X mass. At 2X you can lift a weight with an equal weight, probably using a >1.5X leverage vs. > 1.5X mass to create 2X efficiency. At >2X you ge tvery excpetioanl efficiency, but this only might occur with exceptional design. As a result of this, I conclude that a real perpetual motion device will have about 200% efficiency arising from 1.5 to 2.5 ratio of weight vs. leverage, but all this means is it can lift its own weight.

Miscellaneous: Magnets are never necessary with real perpetual motion.. It does not guarantee high energy return. Rather, energy, return will be directly scalable to mass.

If there is matter without energy, then there is matter from energy. Bad ideas sometimes work or physics does not always work —> perpetual motion might work.

A speed of one unit per unit might be conservative if it defines conservation.

SPECIFIC PERPETUAL MOTION

Friction does not eliminate motion where motion is permitted.

Reactions are possible in a circle, as shown by dominoes, Wheels can turn.

Dominoes can chain-react using higher and higher altitudes. Energy can be created.

Dominoes can accelerate, so friction does not stop everything.

In principle, equilibrium is enough to overcome proportionality problems.

Imbalance can overcome friction.

Equilibrium and imbalance can exist simultaneously through mass-leverage ratios.

With unbalance and a principle of momentum,

PERPETUAL MOTION PHYSICS FOR NON-SKEPTICS

there is no need to lose altitude over time.

All else considered, natural momentum with no net loss of altitude perpetual motion.

With natural momentum and upward and downward motion, potential for return.

Escher Machine : H - V > V (H).

Nathan's Divine Perpetual Motion: 3-d Semaphore + 4-d Wheel.

Natural Torque Device: Natural torque, without net altitude loss, expressed in mass, with recoverability.

First Fully Provable problems: Basically, ,very close to the exact ratios must be kept for it to work ,but the masses are flexible when we keep the ratios. Therefore the counterweight is not really flexible by itself and therefore the leverage ratio must be closely maintained. As a result the mass RATIOS is not flexible, and as a result the lever structure un-weighted must be kept ultra ultra lightweight. These conditions essentially must be met.

First Fully-Provable Master Equation: Given correct ratios, if the balancing weight is flexible, it's in the bag.

First Fully Provable Maximize for Lightweight: long-end lever mass significantly less than 1/2 ball mass.

First Fully Provable Long-end Effective Mass: Total lever mass divided by long-end leverage ratio.

Real Over-Unity Experiment 2: When altitude gain is maximized relative to the slope, the inward spiral can be extreme and still maintain a steep downward drop on return.

PERPETUAL MOTION PHYSICS FOR NON-SKEPTICS

THE SIMPLEST EQUATION

The simplest equation indicating perpetual motion can be found with Newton's law for determining momentum:

$$MOMENTUM = MASS * VELOCITY$$

If this equation holds, then it suggests that

$$VELOCITY = MOMENTUM / MASS$$

Thus, with a little knowledge of trans-finite mathematics in which $1 / 0 =$ infinity, and in which $0 / 1$ is therefore infinitesimal, we get a result in which:

> If MASS > 0, THEN POTENTIALLY, VELOCITY > 0, SINCE MOMENTUM IS A PRODUCT OF MASS AND VELOCITY, VELOCITY CAN THEREFORE BE DERIVED FROM MASS ALONE.

NOTES: In principle, equilibrium is enough to overcome proportionality problems, and imbalance can overcome friction. The two can exist simultaneously through the mass-leverage ratios.

Nathan Coppedge

CONSTANT SPEED PREDICTED FROM UNITS OF CIRCUMFERENCE

1. As the speed is the proportional inverse of the circumference in modular fractional units per unit of motion linear or angular,

2. Speed is constant in objective terms, that is, it cannot exceed one unit per unit at a given time.

3. If it travels 1/4 unit, this requires traveling no more than 4X unit-per-unit of the expected angular momentum.

4. Over 4 units the expected angular momentum is 1/4 unit per 4 units.

5. In both cases, the expected angular momentum is 1/4 unity-unit of time per 1/4 modular unit of the complete cycle.

6. Thus, the momentum tends towards the constant.

7. The only exception to a constant momentum in a functioning machine is if the device constantly accelerates, which would result in a greater overall speed until the functionality of the device was impeded.

AN EQUATION INVOLVING OVER-UNITY

Another equation can be reached that does not prove perpetual motion by itself, but which serves as a standard for certain types of arrangements, in particular the Over-Unity Lever 2.

This equation IS:

IF {THE DIFFERENCE BETWEEN AVG. CHANGE IN ALTITUDE-ENERGY (POSITIVE) AND ZERO} < FRICTION, THEN IF THERE IS UPWARDS MOTION FROM REST AT ANY POINT AND THE CYCLE CAN BE RESET AUTOMATICALLY, THEN THE SYSTEM CAN BE PERPETUAL.

This may seem like a way of making major stipulations, because after all, if a cycle can be reset automatically and it moves upwards from rest, that seems like a perpetual motion machine. But it turns out the additional criteria are important. What they mean is that if friction is the major barrier by a factor greater than the AVERAGE energy occurring during a proven operation, then improving the construction to make it lower friction might result in perpetual motion.

MODIFICATION OF THE HELMHOLTZ EQUATION

Helmholtz free energy states that unobstructed energy with no forces taken into account, can be shown by:

$A = U - TS$

Where A is free or unobstructed energy, U is analogous to internal energy such as momentum or battery power, T is heat-loss, and S is something like global loss divided by scale.

Now modify it for a system in which there is no inputted energy other than altitude, in which heat energy is non-acting, and in which some elements of the system are lifted to a higher altitude with reduced equilibrium --- equilibrium being a force in which motion occurs from rest in terms of altitude, and in which heat energy is also negligible to the function of the system.

Now we have a better definition of free energy! Roughly, for T = 0,

$A^2 = U^X - 2S$,

where X is some unspecified proportional advantage of 2 times or greater for perpetual motion, but no other system. In this equation we can see (hypothetically), not only that beyond an advantage of 2 internal energy competes with entropy, but that beyond 2 the gains are exponential, since U^X begins to take up relatively more of A^2.

Just a theory, but maybe it works!

Note that proportional advantage would be defined in terms of leverage, buoyancy, etc. and as in the other equations, the device must be proven perpetual before the equation works.

THREE-WAY EQUATIONS

Occasionally in my work I've come across difficult cases.

These are cases that only work under very precise conditions.

One example is the Escher Machine (which differs from the real paradoxes introduced by others). It is not to be confused with the M-Motor. The M-Motor is a fraud. The Escher Machine is a difficult, maybe impossible case of simple mechanics.

Another example is the Modular Trough Leverage Device (MTLD). This is a similar, although much easier-to-apply case that might actually work.

The Three-Way equation that emerges with the Escher Machine involves momentum from a backboard, a small resistance from a very slight upward force, and some sort of wedge force occurring on a horizontal dimension, e.g. because the marble is being squenched inwards opposite of its primary motion. According to what I have observed, the squenching motion may be enough to overcome the very slight upwards angle, but only when momentum exists from the backboard. So much for our first three-way equation.

The second Three-Way equation involves proportionality advantages that emerge when two equal weights, one mobile and the other counterweighted, oppose in such a way where the mobile weight is supported along a very slight upward track. It turns out that when the lever is super lightweight relative to the two masses, and the mass of the long end of the lever is compensated in the short end where the counterweight is located, then the lighter mobile mass can be lifted a small distance mostly along the horizontal, until it is able to apply its weight downwards at an unsupported section, lifting the counterweight. Although people with experience in physics have denied this design, there is nothing physically against it, and it has been proven by experiment. That is the second Three-Way equation.

"RATIONAL-ENGINEERING" DEDUCTIONS (REDs):

x: Shape (function) combined with Opposite shape (function)
y : ( ) z : ( ) ...
= maintains potential closed system (function)

example:

(1) lever (function) vs. cyclical track (function),
(2) supported (function) vs. (unsupported) function
(3) short-distance pressure vs. long-distance lift
(4) extended motion (function) vs. contained cycle (function)
(5) momentum (function) vs. momentary activation (function)
(6) stored energy (function) vs. no batteries (function)

= maintains potential closed system (perpetual motion)

ENVOY ---
I hope that this book has given sufficient details of a working model of perpetual motion, and encourage my more
ardent readers to pursue their own work
in this field that has begun so late, but has
so much potential (energy!).

In principia est et infinitum!

Nathan Coppedge

PERPETUAL MOTION PHYSICS FOR NON-SKEPTICS

Nathan Coppedge

OTHER GOOD BOOKS RELATED TO PERPETUAL MOTION:

Nathan Coppedge's Perpetual Motion Machine Designs & Theory

Scientific Papers

Scientific Theories

The Autobiography of the Inventor of Perpetual Motion by Nathan Coppedge

Nathan Coppedge's Designs & Theory and Autobiography Bound Set

BIO:

NATHAN COPPEDGE
Nathan Coppedge is a philosopher, artist, inventor, and poet in some capacity. He is a member of the International Honor Society for Philosophy, and has been quoted on Book Forum and the Hartford Courant. A comment at The Economist cites his possible influence on the economic policy of India. For his work on perpetual motion machines, one website puts him in the ranks of Einstein and Newton. He is also an artist in Hyper-Cubism who has produced over 1000 works. His academic articles span such subjects as objective knowledge, metaphysics, psychology, and immortality. He lives alone in New Haven, CT.